THE WILDLIFE OF
ARABIA

THE WILDLIFE OF
ARABIA

Foreword by
SIR PETER SCOTT

Chairman of the World
Wildlife Fund International

Introduction by
PROFESSOR WILHELM BÜTTIKER

Stacey
International

The Contributors

A.S.T. Professor Abdul Mon'im S. Talhouk (F.R.E.S., London; Ph. D., University of Munich), the distinguished entomologist, is presently adjuncted to the Saudi Arabian Ministry of Agriculture and Water. He has worked as a field entomologist in Saudi Arabia, Lebanon and Syria for over forty years, and was Professor of Applied Entomology at the American University of Beirut from 1953 to 1976. He is the author of a number of scientific papers, and of *Insects and Mites Injurious to Crops in Middle Eastern Countries* (Paul Parey, Hamburg).

W.B. Professor Wilhelm Büttiker (Ph.D. 1921), the eminent zoologist, has spent nineteen years doing research in applied zoology and botany in Saudi Arabia, Africa, Asia and America. He studied at the Swiss Federal Institute of Technology, Zurich and now lectures at the Universities of Jeddah, Riyadh and Berne. His many publications include 130 scientific papers on entomology, parasitology and ornithology, and he is co-editor and co-author of *Fauna of Saudi Arabia* (1979/80, Pro Entomologia, Natural History Museum, Basle).

J.G. and P.R.G. John Gasperetti has lived in Arabia for about thirty years. He is a Patron of the California Academy of Sciences and a Fellow of the Royal Geographical Society. Both he and his wife, Patricia, have contributed several papers on birds, mammals and reptiles to the Journal of the Saudi Arabian Natural History Society, and they plan to prepare for publication an encyclopedic anthology of Arabian herpetofauna, a book on Arabian birds and a study of the natural history of the Makkah By-pass. John Gasperetti is currently associated with Al-Torki Engineers-Constructors in Jeddah.

D.M. Dick Massey, the internationally known photojournalist, received his photographic training at the Rochester Institute of Technology, U.S.A. As a qualified teacher, he has taught design, art and photography at college level. He is an active member of the Biological Photographic Association, for which he has given lectures at conferences and several universities. He has attained recognition in many types of photography but is best known for his award-winning underwater photographs which have appeared in numerous international publications. He has recently turned his attention to cinematography, and has just completed an underwater documentary about a unique group of volcanic islands off the coast of New Zealand. He worked for several years in Saudi Arabia.

Art Director: Richard Kelly
Indexer: Janet Dudley A.L.A.
Map: Ian Stephen

The Wildlife of Arabia

Stacey International
128 Kensington Church Street, London W8 4BH

© 1981 Stacey International

ISBN 0 905 743 27 X

Set in Monophoto Apollo by
SX Composing Limited, Essex, England
Colour origination, printing and binding by
Dai Nippon Printing Company Limited, Tokyo

Acknowledgements

The publishers and authors gratefully acknowledge the very kind help during the preparation of the book of the following:
Dr Nicholas Arnold (British Museum, Natural History); Dr Emilio Balletto (University of Genoa); Frank Courtenay-Thompson; Moira Cuttell (Wildfowl Trust, Slimbridge); Flora and Fauna Preservation Society; Dr David L. Harrison (Harrison Zoological Museum, Sevenoaks); Andrew Harvey (Centre for Overseas Pest Research); Dr D. Hillenius (University of Amsterdam); Michael Jennings; Morrison Johnston (Blackwell and Harrison); Torben and Kiki Larsen; Dr Alan E. Leviton (California Academy of Sciences); Dr Paul Munton (World Wildlife Fund); George Popov (Centre for Overseas Pest Research); Slimbridge Wildfowl Trust; Dr Peter Whitehead (British Museum, Natural History); World Wildlife Fund (International).

The publishers also wish to thank the following who provided the photographs for the book:
Ardea 68 (left: Alan Wearing); Nicholas Arnold 46-7, 54 (bottom), 57 (top), 61 (bottom); Wilhelm Büttiker 7 (both pictures), 8 (map picture 2), 53; Centre for Overseas Pest Research 67 (top: Peter Ward), 70 (bottom: N. D. Jago); Bruce Coleman 10-11 (Jane Burton), 14 (left: Hugh Maynard), 15 (top right: Peter Jackson; bottom right: Joseph van Wormer), 16 (above: Rod Williams), 17 (top: Jan and Des Bartlett; bottom: Cyril Laubscher), 18 (above: M. P. Kahl; left: John Markham), 19 (R. I. M. Campbell), 20 (H. Jungius), 22 (above: Jan and Des Bartlett; left: H. Jungius), 24 (H. Jungius), 25 (above: Jane Burton), 27 (bottom: H. Jungius), 41 (Francisco Erizo), 52 (top: H. Jungius), 60 (above: Allan Power), 62-3 (Stephen Dalton), 65 (bottom left: H. Rivarola; bottom right: Kim Taylor), 66 (above left: Peter Ward), 67 (bottom left: Peter Ward), 69 (Jane Burton), 73 (Jane Burton), 74 (left: S. C. Bisserot), 82 (above: Jon Kenfield), 83 (bottom: Jon Kenfield), 86 (top left: Jon Kenfield), 92 (above: Jon Kenfield), 93 (top and bottom: Jon Kenfield); Robert N. Fryer 33, 42 (above); John and Patricia Gasperetti 14 (above), 32 (above), 28-9, 36 (top and bottom left), 38 (above right), 39, 40 (below), 42 (left), 44 (far left and above), 45 (right), 50 (top and bottom), 51 (above and right), 52 (above left), 54, 55 (top right, middle right, bottom right), 57 (above), 58 (top, above left, above right), 59 (top right, bottom right, below), 61 (top), 65 (top), 75; D.P. Healey 8 (map picture 5), 52 (above right), 56, 66 (top), 68 (above), 74 (above); David Hosking 70 (top); Eric Hosking 26 (above), 27 (top); Michael Jennings 9 (above right and left), 37, 45 (far right); H. Lorenz 76–7, 79, 81, 82 (left), 84–5 (above right), 85 (bottom left and right), 86 (middle left); Dick Massey 79 (top), 83 (top), 84 (above), 86 (bottom left), 87, 88 (above and left), 89, 90-1, 90 (left and top left); MEPhA 8 (map picture 1: R. Turpin; map picture 3; map picture 4: Peter Ryan), 71 (top and bottom right: Torben Bjørn), 72 (above left: Torben Bjørn); Paul Munton front cover, 2 (World Wildlife Fund Photo Library); 66 (centre), 72 (top); Oxford Scientific Films 21 (above: John Chellman); N. R. Phillips 44 (left), 45 (top right); George Popov 9 (left and right), 66 (above right), 67 (bottom right); John Stewart-Smith 23 (above and right), 34 (bottom left and top left), 35, 38 (top), 40 (left), 43 (top right); W. A. Stuart 43 (middle right and bottom right), 38 (above left).

Publisher's Note
All the creatures described or illustrated in this book are referred to by their common English names, where these exist. Scientific names are given in the Index. Measurements, where given, are rounded up or down for convenience.

Title page: A rare photograph of the Arabian Tahr taken in the wild.

Opposite: These illustrations from Sir Peter Scott's Saudi Arabian diaries of 1975 and 1976 show the Redstart, the Racoon Butterflyfish, the Ring-necked Parakeet, the Emperor Angelfish (juvenile and adult), and the Royal Angelfish.

CONTENTS

Ring-necked Parakeet
Psittacula krameri.

British Embassy Residence
garden.
Most numerous in November 1975
on seeds of Caesalpina pulcherrima
Peacock Tree.

Pomacanthus imperator
Emperor Angelfish.

Juvenile.

Adult.

Pygoplites diacanthus.
Regal Angelfish.
Red Sea. Saudi Arabia.

FOREWORD

by SIR PETER SCOTT
Chairman of the World Wildlife Fund International

They say that travel broadens the mind. It is certainly very enjoyable, especially if you have been trained as a painter and a naturalist, and have a wife who likes it too. It has been our great good fortune to travel in the Arabian peninsula eight times since 1974.

The vast deserts of that region support an impressive number of animal and plant species that are beautifully adapted to survive in an arid climate. The Arabian Oryx had been exterminated in its wild range though it bred in captivity and has now been re-introduced to the peninsula; and at least one of the three species of gazelles is in danger of extinction. But not all of Arabia is desert. In the south the mountains rise to over 3,000 metres(10,000 feet) and there are extensive forests along the escarpment which runs up the southwestern side of the peninsula. In the Asir in Saudi Arabia a fine National Park has been established by the Governor of the Province. Around the coasts there are sand beaches, up some of which marine turtles still come to lay their eggs at night. There are fringing reefs with their brilliant profusion of corals and reef fish, and there are a host of islands, some of them with little-studied colonies of nesting sea birds. Among the islands there are still some remnant populations of Dugongs – the original 'mermaids'. Twice a year vast numbers of migratory birds sweep across Arabia – southwards in the autumn, and back to their northern breeding grounds in the spring.

From our visits to Arabian countries we have acquired a host of memorable wildlife experiences, such as landing on a small island just off the coast near Jeddah when the spring migration was at its height. The tiny bushes dotted about gave sparse cover, yet every one was full of birds – warblers, redstarts, rock thrushes, shrikes and Corn Crakes. I was able to approach within about half a metre of a Great Reed Warbler which kept perfectly still believing I had not seen it, and I was able to retire without disillusioning it. Once, on the Jebel Aswad in Oman, I looked over a near-vertical cliff of about 600 metres(2,000 feet) high, and saw, quite close across a cleft, two of the very rare Arabian Tahr – a mother and kid – walking slowly along a narrow ledge in full sunlight. From the top of the escarpment in the Asir we watched vultures and a Tawny Eagle soaring close in front of us, and looked down on the courtship of Tristram's Grackles on the cliff face below. Arabian gardens attract migrant birds and the lawn of the British Embassy in Jeddah often teamed with Yellow Wagtails, shrikes, parakeets, and almost always Hoopoes. One day when I had been watching a Wryneck, I heard cranes calling – Demoiselle Cranes – and presently three hundred of them passed directly over me as I lay on my back on the lawn looking up at them through binoculars.

But our most vivid and colourful memories are of diving and snorkelling on the coral reefs – finding the Red Sea species of the Racoon Butterflyfish and a profusion of Emperor and Regal Angelfish, swimming among harmless sharks over an exquisite reef off Muscat, and with friendly sea snakes off the coast of Qatar. A very special memory involves visiting a famous captive herd of Arabian Oryx in Qatar and finding there the caterpillars of one of the most beautiful moths in the world – the glorious Oleander Hawk Moth – which we subsequently bred in captivity to the third generation.

Arabia is still rich in its wildlife but conservation measures are urgently needed in many areas. Perhaps collaboration between the countries concerned, and consultation with the International Union for Conservation of Nature and Natural Resources, and the World Wildlife Fund, might lead to an International Conservation Convention for the region. And as is the case everywhere in the world, there is always need for more education and greater public awareness of what will be lost if practical conservation is not more widely supported.

In former times mankind lived in a more harmonious relationship with nature in Arabia. It is written in the Holy Koran (*Sura 6.39*): 'No kind of beast is there on earth nor fowl that flies with its wings but is a folk like you . . .'. Those 'folk', with whom we share this planet, are worthy of our help to survive in this difficult century.

This illustration of the Hoopoe is taken from Sir Peter Scott's diary of his visit to Saudi Arabia in 1975.

INTRODUCTION
How Arabia's animals came to be there

by PROFESSOR WILHELM BÜTTIKER

Some twenty to thirty million years ago the Arabian peninsula was joined to northeast Africa. During this period the African Rift Valley system, from the Jordan valley to central Africa, was formed. It was at this time also that the Red Sea began to take shape – as an inland lake, initially, until the rift at the Horn of Africa was wide and deep enough to link it to the Arabian Sea and the Indian Ocean. Following this development came an influx of marine animals to the newly-formed sea.

A more ancient arm of Arabian-Indian waters is the Arabian Gulf where similar conditions prevailed. Some seven hundred million years ago the Gulf's extent was considerably larger; its western shorelines were closer to the ancient Arabian Shield. There is also strong evidence from many geological periods that much of northern, central and southern Arabia was covered by seas for a long period. The proof is the tremendous wealth of marine fossils that have been found in parts of the peninsula. They include clams, shells, corals and urchins, particularly from the Jurassic and early Cretaceous geological periods of between a hundred and eighty and a hundred and sixty million years ago.

A steady uplift of the peninsula took place, hinged at its eastern edge. Gradually the landmasses became extended and at one time, during the later Cretaceous period, approximately seventy million years ago, there appears to have been dense coastal vegetation. The petrified forests of large trees still found in some places are the witnesses of this lush growth. This was also the age of the giant reptiles. Rivers ran west to east, emptying into a mighty river that is today the Gulf basin. There certainly existed several million years ago a land bridge across the mouth of today's Gulf which allowed access of animals of Oriental origin, some of whose descendants (for example the Tahr) survive today.

In the Tertiary period, a new type of fauna appeared: marsupial mammals. Their existence is also confirmed by the presence of fossils. Later came other mammals closely related to the species currently found in the eastern region of the peninsula. The further east we search for fossilized animals, the more recent the geological sediments we strike. Near the Arabian shoreline of the Gulf we find sea sediments and fossils of the last few thousand years, again in the form of shells, clams and corals.

Arabia's geological story makes the subcontinent outstandingly interesting to the zoologist; it is the meeting-point of three major zoogeographical regions: the Palaearctic (today's Europe, North Africa and northern Asia), the Afrotropical and the Oriental. The influence of each is strongly felt in the animals of the peninsula. The evolutionary link with Africa is particularly strong: many African animals including the hyaena, baboon and leopard, as well as hundreds of species of snails, birds, insects, fish and reptiles are elements of African fauna found in Arabia.

The influx of Palaearctic animals (such as the wolf) into the peninsula occurred particularly during cooler climatic periods, the Ice Ages. Of the Oriental animal groups many found their way to Arabia across the land bridge at the Strait of Hormuz. For long periods the Arabian peninsula, bridging Africa with Asia, was thus well positioned for the migration of terrestrial animals. However, in the intermittent arid epochs, the same subcontinent became a barrier to the interchange of species because of the huge barren land surfaces of stony and sandy desert that had to be crossed.

Below left: The fascinating 'instant crustacean', *Triops granarius*, lies dormant for years until the arrival of a rainstorm when it moults from its cyst and grows to its full size (shown here).

Below right: These petrified tree remains prove the existence of coastal forests in the mid-Cretaceous period.

ARABIAN PENINSULA

LEBANON
SYRIA
JORDAN
IRAQ
BADANAH
KUWAIT
IRAN

Jawf
Tabuk
NAFUD
Midian
Hail
QASIM
Buraydah
DAHNA
NEJD
Jubail
Dammam
QATAR
ARABIAN GULF
Musandam
Sharjah
Dubai
GULF OF OMA
J. Hajjar
Medina
HEJAZ
Yanbu
HEJAZ
Hofuf
DOHA
ABU DHABI
U. A. E.
Batinah
J. Akhdar
MUSCA

RED SEA
Jeddah
MAKKAH
Taif
SAUDI
J. Tuwaiq
ARABIA
Wahiba
Sands

SUDAN
ASIR
Tihama
Abha
Najran
RUB AL KHALI
Salalah
OMAN

Farasan
Islands
Jizan
NORTH
YEMEN
SAN'A
SOUTH YEMEN
HADRAMAWT
Mukalla

ARABIAN SEA

N

0 100 200 300 400 500
KILOMETRES

ADEN

ETHIOPIA
SOMALIA

Opposite: The varied habitats of Arabia's wildlife include:

1 Sand deserts, which support negligible life.

2 Wadi beds, a few of which retain pools of water year-round.

3 Fertile plains and mountains in the Asir.

4 Scrub desert and rock outcrops.

5 Sabkha salt flats, often along the coast.

Little is known in detail about these zoogeographical conditions but field research in progress will uncover a wealth of information.

The severe climate of Arabia has produced among certain animals extraordinary solutions to the problem of the survival of the species. Perhaps the most remarkable case of all is that of the 'instant crustacea'. Observers have sometimes been astonished to see small shrimp-like creatures swimming about in pools that appear in the desert for a few days after the rare rains – sometimes in spots where no water whatsoever has lain for seven years or more. Where have these animals come from? Zoologists have uncovered the phenomenal life-cycle of this crustacean, the *Triops granarius* (shown on page 7), which is presumably a mutated survivor from times long ago when Arabia was plentifully watered. The *Triops granarius'* egg, laid by the female as the momentary desert pools dry up, immediately develops into a hard cyst, which sinks among the grains of the sand and remains there, close to the surface, year-in, year-out, quite dormant, unchanging yet imperceptibly

alive, until another rainstorm fills the little depression. Then, with great swiftness, the embryo 'moults' from its cyst, lives its brief life in the shallow pool, grows to its full size, and lays or fertilizes eggs, which then repeat the cycle.

Other intriguing clues to the history of Arabian fauna are provided by the rock graffiti found in many areas of Saudi Arabia. These depictions on granitic and sandstone rock give evidence of the fauna that was present there some four to five thousand years ago. Large areas of the peninsula were steppes and savannah then, the result of higher rainfall and cooler climatic conditions. According to the graffiti, a few examples of which are illustrated here, man of this period was hunting ostriches, gazelles, ibex, wild oxen and camels by means of spear, bow and arrow, and dagger. Many images also show in abstract form lions, wolves and other canines, as well as humans in an adorant position.

Since the coming of firearms, the devastation of Arabia's wildlife has been grave indeed. But today, under the enthusiasm of the heads of State of almost all the countries, and particularly of Saudi Arabia and Oman, determined efforts are being made to conserve and to educate the people in the importance of the natural heritage of one of the most valuable regions, zoologically speaking, of the whole world. Already some heartening successes have been scored.

W. Büttiker

Above left: Wild oxen were once present in the Arabian peninsula, but, like the Ostrich, they no longer exist there.

Above right: Rock graffitis show that gazelles were the target of the hunter in earlier times.

Left: This rock graffiti in the Asir depicts a pair of Arabian Ostriches, animals now sadly extinct in the peninsula.

Right: The Ibex, as is shown in this rock art, was one of several animals hunted by man in the past.

MAMMALS

What mammals, the casual observer might be excused for wondering, can exist in a territory like greater Arabia. Mammals, like humans, need to shelter from the heat and the cold; they need water; they need a dependable source of food – either other animals, or vegetation. Yet one might cross the Arabian subcontinent from the Red Sea to the Gulf at any time in eleven months of the year and be convinced that no mammal species – at least, no mammal of any size – could endure the climatic rigours and the absence of food sources.

Yet, this surface impression would be most mistaken. For Arabia, historically – and sometimes by the merest thread today – sustains a surprising variety of mammal species, some of them quite large animals and several unique to the peninsula. This is due partly to the deceptive reputation of Arabia as a terrain of waterless desert, and partly to the extraordinary powers of evolving species to adapt in their bodily mechanisms and lifestyles to a hostile climate.

Perhaps the most dramatic story of the recent rescue of an important species of large mammal is that of the Arabian Oryx. And so let us begin with this remarkable species, and its fellow gazelles. The Arabian Oryx (illustrated on page 22), evolved like many Arabian species from relatives in tropical Africa, became an early victim of the gun in the hands of hunters and tribesmen. The terrain it occupied, the very sparsely vegetated rim of the true sand desert, could never support large herds. By the early 1960s zoologists knew that the Arabian Oryx was on the brink of extinction. Perhaps with no more than a year or two to spare, several Arabian Oryx were captured alive by zoologists and transported, first to Kenya, then to Arizona. From the early 1960s not a single sighting of another Oryx was made in Arabia. But from these tiny beginnings, a herd grew and flourished in the Arizona desert; and in 1980, nearly two decades later, a group of selected animals from this herd was re-introduced into the wild in Arabia, in the Wahiba sands in central Oman. The story of the Oryx is a beacon of hope to those fixed upon preserving the wildlife heritage of Arabia for future generations.

The Oryx, standing 100cm(39in) at the shoulder, is the largest of the cloven hoofed animals of Arabia. The Idhmi gazelle, for example, stands about 60cm(24in) – a long-legged graceful creature, red-brown on the back with white underparts; it is shown on page 22. Once it was widespread on the coastal lowlands, and adjacent foothills, and on some inshore islands, but it is now extremely rare. Its fellow gazelle, the Afri, is very slightly shorter. Sandy brown, with a black tail and straight horns, it was also once common except in coastal areas. Its endemic Arabian sub-species, the *Gazella dorcas saudiya*, is alas probably extinct through overhunting. Largest of the Arabian gazelles is the Rheem, of which some appear almost white. This lovely animal (shown on page 23) with its lyre-shaped horns is known to survive on the edge of the Rub al Khali.

Preceding pages: Extending its superbly long and striped bushy tail, the Common Genet lies along a branch, its bright eyes glinting. Its extremely rare Arabian relation lives in the hills and mountains of the peninsula.

The Oryx and the gazelle have had few enemies besides man; they are either too fast or too powerful. The wild sheep and goats of Arabia, and the Ibex, are more vulnerable. Once the Ibex was widespread: today it is known to exist only in a few areas including the mountains of the previously volcanic Northwest. A darkish fawn, with pale underparts and dark and light patterns on the legs, the male is known for its prominent black beard, long ears and scimitar-shaped horns. A female Ibex is illustrated on page 21. Arabia's Wild Sheep is also equipped with robust horns which tend to curve into a circle – although females are sometimes without horns. It is still reported from time to time, but its distribution is largely a mystery. Little is known for certain today of the precise whereabouts of the Wild Goat. Sometimes this grey or brownish-grey animal can reach a height of 94cm(36in). The male possesses a prominent dark beard, a black stripe across its shoulders, and a black crest down the top of its back.

More is known of the Arabian Tahr, (pictured on pages 2 and 20), the small wild goat found only in the mountains of northern Oman. This extremely shy animal has perhaps survived because of the inaccessibility of the territory it occupies – steep cliffs and gullies which are impossibly dangerous to man. The Tahr ranges this forbidding habitat, surviving on wild fruits, seeds, shoots of shrubs and grass. But unlike other large mammal species, such as the Oryx, the Tahr requires water daily.

Until quite recently in evolutionary history, the enemy of the Tahr would have been the Leopard, a sub species of which, *Panthera pardus nimr*, is still known to survive in the most remote areas of the southern and western mountains of the Arabian peninsula, and on the extreme tip of territory, the Musandam Peninsula, at the entrance to the Gulf, where a Leopard carcase was recently found. The Arabian Leopard is almost white in its ground colour, with brownish-black to jet black spots and rosettes. It has been mercilessly hunted for its pelt.

The Cheetah, by contrast, occupies not the rocky mountains and outcrops, but the desert steppes, depending for its hunting success on its speed. Possibly the Cheetah is extinct in Arabia, although a specimen was taken in Oman in 1977. Its close African relation is pictured on page 18. The threat to the big cat is of course not only man's bullet directed at it, but also the elimination by man of the mammals it survives on.

Like the Cheetah, the Caracal, with its striking ear tufts, was once widespread but rare today. With a length of 86cm (33in) nose to tail, the Caracal, shown on page 19, is much larger than the Sand Cat, shown on page 18. The latter lovely animal, with its broad ears, pale isabelline colour, and feet covered with hairs to aid movement in soft sand, is also rarely seen in the wild. Its relative, the Wild Cat, is the ancestor of some domestic cats. Ashy grey in colour, with fine speckling, and broad triangular ears, it is most immediately identified by its black tipped and ringed bushy tail; but like most cats it moves and hunts at night and is seldom seen except by the determined naturalist.

There is a wide variety of smaller mammals for the lesser

cats to live on – such as the Rock Hyrax (the Coney of the Bible) which is found in the mountainous areas of the West and South and in the central mountains also. This scurrying animal of about 50cm(20in) long, with no tail, large black eyes and rounded ears and a peculiar nocturnal mewing wail, is illustrated on page 17. Its blunt toes resemble tiny hooves, and its skeleton proves its common ancestry with the elephant and the rhino. The Hyrax lives in colonies feeding on leaves, roots, bulbs, berries and seeds.

In the mountains of the Southwest, where water is easily dependable, several other small mammals exist. The Genet on page 10, for example, is a slender, cat-like nocturnal predator, which preys on rodents, lizards or snakes. Two species of mongoose live in Arabia, the Indian Grey Mongoose, shown on page 16, and the White-tailed Mongoose. The weasel family is represented by two species. The Ratel, or Honey Badger, pictured on page 17, was once common but it is now very rarely encountered. A heavily built nocturnal animal, it is covered with coarse black hair except for a prominent white mantle down its back. It survives by burrowing its prey out of their refuges – little rodents, lizards or snakes. The Stone Marten, by contrast, is a more slender and graceful animal, with a bushy tail, round face and pointed muzzle; it must be in reach of water for it almost certainly requires frogs and fish to keep it going.

The open country is, for the most part, the habitat of Arabia's dogs. Jackals, foxes and wolves are common, though not often seen. The Arabian Wolf, the age-old enemy of the pastoral tribes, is widespread; it resembles a German Shepherd or police dog. Even the wolf's survival is now threatened, and it is a protected species. Like the wolf, but smaller, is the Jackal (illustrated on page 15), found in the oases of eastern Arabia. The Common Fox and Rüppell's Fox are widespread though the latter is rare. The Fennec, a rare denizen of the sandy wastes, is, as can be seen on page 15, a little animal of a mere 58cm(22in) from nose to tail with large ears and eyes. The Striped Hyaena, with its hairy dorsal mane and striped legs, is now restricted to rough jebel areas; its ferocious appearance belies a timid nature.

Perhaps the animals the most effectively defended from all predators are the hedgehogs and the porcupine. Though they are not related, both these species depend on their sharp spiky spines, which confront an enemy from every direction, to protect them. Hedgehogs, of which there are three or four species in Arabia (see page 14), are relatively abundant insectivores. The Porcupine, a large rodent, whose tail is equipped with a group of hollow quills rattled with startling effect in moments of annoyance, is today extremely rare. It is a nocturnal vegetarian, known to devastate gardens.

Arabia does not have trees in sufficient density to support monkeys. But in the hills and mountains of the Hejaz and Asir are to be found troops of *Hamadryas* Baboons, such as that shown on page 14, congregating at waterholes and in places where they can reach the refuse of cities – for they are at ease in the proximity of man and appear to survive in gleaming health on what man throws out. The baboon (like man himself) has the peculiar advantage of being able to take nourishment from a wide variety of food.

Besides baboons, the mammals that have the best chance of survival in Arabia today are the smallest animals. There are perhaps six species of shrew to be found, usually grey in colour with soft dense fur; they live on insects of every sort. The peninsula is also the home of at least twenty-five species of small rodent, several of which are illustrated on the following pages – the Jerboa for example, which moves in dramatic leaps on strong hind legs, balanced by its white-tufted tail that comprises sixty per cent of its length. The five-toed *Allactaga euphratica* Jerboa is found in the Northeast while the Common Lesser Jerboa, which is shown on page 26, is widespread.

Dormice and House Mice are also present as is the widespread Spiny Mouse, which is shown on page 25, and its relative the Golden Spiny Mouse, with a tail that breaks off easily as a defense mechanism – leaving the predator without the meal he had anticipated. There are eight species of gerbil – a mouse-like animal, with soft fur of a sandy colour, two examples of which are shown on page 27. The Bushy-Tailed Jird, some 27cm(10in) long (of which more than half is tail) and thickly haired in a pale sandy colour with white underparts, is found from Riyadh to the Northwest. The Fat Jird, reddish-clay in colour, sports a blackish tuft. Of the rats, the Black Rat is the most common, although the Brown Rat is the more formidable.

Perhaps the most mysterious of Arabia's animals in their habits are the bats. In the extreme Southwest is found the *Eidolon helvum aegyptiacus* and in the West and in Oman the *Rousettus aegyptiacus*. Both possess large eyes and broad membranous wings spanning more than half a metre and live on fruit. There are also several insectivorous bats which inhabit the innumerable caves of Arabia, skimming about the rock crevices and thick foliage of the trees at night, when all their feeding is done.

Arabia can make its claim to several marine mammals – particularly the remarkable Dugong, reputedly the inspiration of the mermaid legends. The Dugong has an impressive distribution in the oceans of the world, but in most of its range it has been mercilessly slaughtered for its flesh, hide and tusks. Quite unlike the mermaid in reality, it is a vast, ugly torpedo-shaped creature. A unique species survives in the Red Sea – *Dugong dugong tabernaculi*. The Red Sea is also the home of Bryde's Whale and the Red Sea Bottle-nosed Dolphin, which, if it shares the intelligence now widely attributed to dolphins, must surely ask itself when the time will come for man to decide whether he cares about preserving his natural heritage.

J.G., P.R.G. & S.I.

Above: Colonies of baboons (*Papio hamadryas*) are content to live in proximity to man, but adult males can be tetchy and aggressive, and bark harshly or roar. Baboons appear to have been venerated in Ancient Egypt, where they were sometimes mummified.

Left: The three or four species of hedgehog known in Arabia are nocturnal animals. This Desert (or Ethiopian) Hedgehog's sharp spines, banded with black and white, protect it from its enemies when it rolls itself up.

Top right: During the dry season, the scavenging Jackal is forced to look for food in cultivated areas where it devours watermelons and pumpkins. It hunts in packs at dusk and by night when its dismal howl may be heard.

Bottom right: A species of fox extremely rare in Arabia is the Fennec. It is a small animal with a sharp muzzle and pointed ears which lives in the sand wastes: here its close African relation is shown.

Above: Perched on a rock, this Indian Grey Mongoose, with its pointed and alert face, is an agile climber. It sometimes inhabits hollow tree trunks and it has a taste for snakes, small animals, certain fruits and roots.

Opposite top: The Ratel is a proficient and speedy climber, able to reach the top of a telegraph pole in four seconds. It lives on rodents, birds, frogs and insects which it catches at night, and when alarmed it will burrow for cover in a few minutes.

Opposite bottom: Despite its somewhat clumsy appearance, the herbivorous Rock Hyrax is a master of agility. Shy and retiring, it lives in colonies in deep rock crevices, from where it appears in the early morning and evening. Its skeleton and foot formation prove a common ancestry with the elephant and rhino.

Above: The athletic Cheetah is superbly adapted to fast travelling. Its prey includes gazelles and smaller desert mammals, whose decline may have made the Cheetah extinct in Arabia – sightings have not been recorded for a few years. It is extremely similar to the African example shown here.

Left: An inhabitant of arid terrain, the Sand Cat can survive on very little water. Its hairy feet make gripping easier when moving on soft sand, while its low-lying ears possibly help it to flatten itself when stalking prey.

Opposite: The elegant and secretive Caracal is a fast mover and a great jumper; it catches with ease birds which fly within reach.

Opposite: The Arabian Tahr, a wild goat, is unique to the mountains of northern Oman. Entering Arabia by the former land bridge with southern Persia, it has developed there in isolation. Rarely seen, it lives on the craggiest of mountain precipices.

Above: The Ibex lives in the rocky mountains of the Northwest; it has a long and fascinating history. Recorded in both palaeolithic and neolithic rock art in the Jebel Tubaiq, it was believed to have been symbolic of the moon god in the days of the Queen of Sheba.

Above: The once abundant beautiful Arabian Oryx was successfully re-introduced into Oman in 1980 having been extinct in the wild since 1962. It could exist almost without water for months on end, but could not escape the gun. It is thought to be the origin of the legend of the unicorn.

Left: Once widespread, but now extremely rare, the long-legged Arabian gazelle known as the Idhmi lives in the mountains, foothills and coastal plains of the southern and western areas of the peninsula. Travelling agilely over the rough ground, it moves in graceful leaps.

Above: The largest Arabian gazelle, the Rheem, runs with its neck outstretched at an astonishing pace, but it does not leap or bound. It congregates in herds of up to a hundred which run packed tightly together. Its numbers in Arabia are now greatly diminished. *Right:* Two Rheem calves.

Above: The nocturnal Spiny Mouse is widespread in rocky habitats throughout the peninsula. Its easily broken tail helps it to escape from predators, which are left with just the end of the tail.

Opposite: One of the most versatile mammals of the peninsula is the soft and furry Hare (*Lepus capensis omanensis*). It occurs throughout Arabia and is a mainly nocturnal creature. Though hunted by man, wild cats, wolves and foxes, it survives in considerable numbers.

Above: The Jerboa is a particularly adaptable colonist of the deserts of Arabia, just as at home in sand as in stony steppes. A vegetarian, with a lengthy tail, it can jump from a standing position up to two thirds of a metre in the air.

Opposite top: A strictly nocturnal and retiring creature is the Gerbil. Living in the rocky steppes of the desert, it feeds on seeds and frequents small holes under slabs of stone.

Opposite bottom: The Jird is a robust rat-like gerbil. It is also known as the Sand Rat, though its habitat ranges from desert and steppe land to agricultural terrain. Its soft fur extends along its tail, which has a tuft on its tip.

BIRDS

A surprising variety of birds – more than five hundred species – have been recorded in Arabia, including lordly birds of prey, exotically plumaged species, elegant waterside birds, waders, doves, larks, thrushes and many more. About a dozen or so belong only to the area but the peninsula reaps the reward of being a zoological crossroads, of lying across migration routes, and of providing varied habitats.

Arabia nestles between three major zoogeographical regions of the Old World: the Palaearctic, the Afrotropical and the Oriental. The influence of the avifaunas of each is felt to a greater or lesser degree in all parts of the peninsula. Its northern and central sectors are in the arid Eremian zone of the Palaearctic region which embraces Europe, North Africa and northern Asia; most birds resident in Arabia belong to this zoogeographical region. Others belong to the zone covering Africa south of the Sahara – especially those found in the Tihama coastlands and the middle altitudes of the western mountains, north to Taif. Typical among these African birds are the Little Grey Hornbill, the White-browed Coucal and the Dikkop. Along the Gulf are found several species from the Oriental region, which includes the Indian subcontinent, southeastern China and Malaysia. Among Indian birds in Arabia are numbered the Indian Roller, the tiny Purple Sunbird, and the White-cheeked Bulbul.

Arabia's position athwart the routes used by many species in their twice yearly migrations provides a bonus to the variety and numbers of its birds: each autumn some two to three thousand million birds of at least two hundred species make their way to warmer climes, either passing over Arabia or stopping to winter there.

The outward journeys take place between August and October each year. The majority travel from Eurasia to Africa, but others more from Europe and the Middle East to India. Yet others fly from India to Africa. The return journeys occur between February and April. Among these migrants are rollers, bee-eaters, kingfishers, wheatears, warblers, swallows, swifts, martins, redstarts, wagtails and pipits, shore birds, doves, Quail and cranes. The Houbara Bustard, as well as many ducks and even occasionally geese and swans, terminate their migration in Arabia to winter there.

This winter influx also includes a variety of birds of prey and vultures. Of these, the Saker Falcon is sometimes caught alive in ingenious traps bated with pigeons, and within a short period of ten days to two weeks, trained to catch hares and Houbara. Arabia's birds of prey also include harriers, buzzards and several species of eagle. The Kestrel and the Barbary Falcon are resident breeders, although they are also partial migrants. Of the vultures, the Black Vulture is another winter visitor, though it may just possibly be resident, as may the Lappet-faced Vulture found in central

Preceding pages: The remarkable Hammerkop is a relative of the herons. It inhabits the streams and pools of the Tihama and Hejaz and the Asir mountains and is generally a solitary species. In winter it can however sometimes be found in groups in sheltered lowland watercourses.

Arabia in winter. Rüppell's Vulture in the western areas and the Egyptian Vulture are to be seen the whole year round, as is the huge Griffon Vulture, which is illustrated on page 36. The stately Lammergeyer nests in the higher more remote mountain areas.

Of the other birds of prey, the owls are rather a mysterious group. Some such as the Long-eared and the Short-eared Owl are visitors, but the majority, like the rare Hume's Tawny Owl, illustrated on page 37, are resident breeders. The owls are a varied family and include two large species, the Eagle Owl and the Spotted Eagle Owl, the diminutive Scop's Owl, the Barn Owl and the Little Owl.

Some resident birds have complicated patterns of migration within the peninsula, depending on the erratic food resources of the arid regions. Their movements are also dictated by the availability of surface water, always unreliable. The Rufous Bush Chat, another summer visitor to Arabia, breeds in the east of the peninsula, while the Arabian Babbler and the Black Bush Chat are residents which appear to take part in local movements. These three birds all frequent bushy thickets and gardens.

The tremendous diversity of natural habitats in Arabia contributes to the variety of birdlife. Mangrove swamps, cliff faces or sand beaches bordering rich tropical seas, coastal lowlands, high craggy mountains, terraced agricultural land and rocky desert steppes: all these types of terrain exist in the country and all have their different inhabitants.

On the desert steppes and in the arid wastes several birds are resident – particularly larks. The most widespread is the Hoopoe Lark, a true desert bird often seen far from water. The Black-crowned Finch Lark is also frequently found in the deserts as are the Crested Lark and the Desert Lark. Temminck's Horned Lark is common in both the North and the Northeast. The large flocks of Short-toed Lark are winter visitors.

Another typical bird of the desert areas is the Cream-coloured Courser, a nomadic resident. In the rock desert lives the Arabian See-see Partridge, a bird that is most frequently seen in the western and southern foothills. Also in the desert areas, near water sources, may be found six species of sandgrouse, birds which are closely related to the pigeons. Although they dwell mainly in the desert, large flocks of them daily fly great distances to water. The Coronetted Sandgrouse, the Spotted Sandgrouse and the rarer Pin-tailed Sandgrouse live in northern and eastern Arabia; the southwestern, southern and southeastern sectors are the homes of Lichtenstein's Sandgrouse and the Chestnut-bellied Sandgrouse.

The Brown-necked Raven can exist far from water in even the most remote deserts. It is, however, found throughout the land, and is especially abundant in farmed areas and in the mountainous reaches that are also the habitat of the Fan-tailed Raven. Both these species are found in such large numbers that they are considered pests.

The most remarkable bird population exists in the Southwest of the peninsula – to the south of Makkah and Taif, along the coastal stretches to the heights of the outstanding escarpment that reaches altitudes of over 3000 metres(10,000 feet), and on to the Yemen, where the mountain masses continue to near the Gulf of Aden and eastwards to Dhofar.

In the wooded lowlands of these western and southern sectors, the resident birds number among them the Black-capped Bulbul, Rüppell's Weaver (shown on page 44), the Nile Valley Sunbird, the Palestine Sunbird, the Shining Sunbird and the Little Green Bee-eater (shown on page 43). The Namaqua Dove is found eastwards towards central Arabia, while the Palm Dove and the Eurasian Collared Dove (shown on page 42) are to be seen in the northern and eastern parts of the area.

In the highlands, in the vicinity of Abha, there is a relict population of the Magpie which occupies a very restricted area. Among the other relicts (species left behind in the temperate mountains of southwest Arabia when the ice sheets further north retreated, and which now are completely isolated from their nearest relatives) are the Yemen Linnet, the Arabian Woodpecker, the Yemen Thrush, the Yellow-rumped Serin, the Tristram's Grackle, which extends north to the Dead Sea, the Arabian Accentor and the Golden-winged Grosbeak.

As was previously mentioned, many of the other birds found in this southwest sector have close affinities with African species – the Amethyst Starling, the Red-eyed Dove and the African Silverbill (illustrated on page 45) are further examples. Where there is water close by, Abdim's Stork, the Hammerkop (see page 28) and the Grey-headed Kingfisher may be found. Some of these same species are also found in the brush lands, acacia forests and farmlands or the Red Sea littoral.

The western coastal lowlands are also the home of a most remarkable bird, the majestic Arabian Bustard. Alas, its survival is threatened by human nest robbers just as is the Arabian Tufted Guineafowl's; this bird, illustrated on page 39, is currently considered by taxonomists to be a race of the African Tufted Guineafowl, though further study may prove the Arabian bird to be a distinct species. Overhead fly great flocks of the Arabian Golden Sparrow, shown on page 44, and Bruce's Green Pigeon, a bird that lives mostly on wild figs. The charming little Harlequin Quail and the brilliantly coloured Abyssinian Roller, shown on page 42, are also confined to the Southwest.

Among the many migrant summer visitors to the area is the lovely White-throated Bee-eater. On occasion the stately Lammergeyer may be seen overhead or the strikingly marked eagle, the Bateleur, circling the cliff faces of steep escarpments like that on the approach to Taif. This last creature is one of the world's most colourful eagles: it has a red beak and red talons, a reddish back, white breast, black and white underwings and purplish-black wing coverts and head.

Bounded by tropical seas, Arabia's coasts are a rich habitat for water birds. Storks, ibises, Spoonbills, herons, egrets, bitterns, plovers, stilts are all encountered, and many more. The Red Sea coast is the home of two interesting gulls, the White-eyed Gull, which can be seen on page 40,

and Hemprich's or Sooty Gull. Their range extends across the coasts of southern Arabia and East Africa: they are endemic to the northwestern Indian Ocean. There are also many Flamingo present throughout most of the year, and occasionally pelicans are seen. The Pink-backed Pelican is a resident breeding bird in small numbers in the mangroves of the Red Sea. The Flamingo can be seen on page 34 and the Pink-backed Pelican on page 32.

In the southern sea coast areas an occasional Red-billed Tropicbird (illustrated on page 41) may be sighted at its nest on islands in the Gulf of Aden. On the Arabian Gulf coast one of the most common birds is the endemic Socotra Cormorant, found in large nesting colonies on some of the Gulf islands. It is illustrated on page 35.

Inland, the continuing process of the development of water resources in the form of increased irrigation, water storage and distribution is creating more suitable habitats for many species. In the cities, towns and villages the tapping of deep aquifers is providing an abundance of water never known before; garden greenery, grain crops and other vegetation are consequently increasing. As a result, many birds such as the Rose-ringed Parakeet, the Black Kite and the House Sparrow are propagating and increasing their range dramatically especially around Jeddah, Riyadh and in the eastern towns. In Oman the Indian Crow is doing the same. The Hoopoe is widespread during the cooler months, particularly in the higher mountain areas of the West where it is resident and breeds. More than any bird, it has achieved fame in story and legend and is mentioned in the Holy Koran as the messenger of King Solomon.

Since the advent of increased water resources and the decrease in trapping and hunting practices, some birds whose numbers had been on the decline are now assured a lasting existence. Sadly, one of the casualties before the adoption of conservationist attitudes, was the Arabian Ostrich, (see page 9), which was native to the desert steppes and adjoining sands. The adult bird was the hunter's greatest prize, while its young and eggs were eagerly sought for food. This assault eventually constricted the Ostrich's distribution to the waterless wastes; the breechloading gun and the motor car with sand tyres sealed its fate, and about forty years ago it became extinct.

J.G., P.R.G. & S.I.

Above: Birds visiting southwestern Arabia from Africa include the impressive Pink-backed Pelican. In recent years it has also been found breeding on the southern Red Sea coast of Saudi Arabia.

Opposite: Always thought of as an uncommon migrant and visitor to Arabian shores, the Spoonbill was found breeding in mangrove swamps near Kunfuda on the Red Sea in 1980. The yellowish patch on its neck is present only in the breeding adult.

Top left: The graceful Flamingo visits the Arabian coasts mainly in winter. The species once bred in the north of the Gulf, and might return to this breeding ground.

Bottom left: A particularly majestic visitor to the shores of the Gulf is the Great White Egret. Like the several other herons and egrets of Arabia it moves northwards in spring to breed in Asia Minor, Mesopotamia and the Caspian region.

Opposite: The only endemic seabird of the Arabian region is the Socotra Cormorant. Vast, densely packed colonies breed on islands in the Gulf.

Top left: The Osprey breeds on all the shores of Arabia. It dives into the water after its prey, using its particularly effective talons to catch and hold the fish on which it feeds.

Bottom left: A camel dies and a Griffon Vulture feeds. This scavenger is familiar in the western half of Arabia but quite rare in the Gulf region. Six species of vulture occur in Saudi Arabia.

Opposite: Acknowledged as one of the world's rarest owls, the Hume's Tawny Owl was discovered in 1975 as a resident in the Tuwaiq Escarpment in central Arabia.

Top: The Lesser Golden Plover comes to the Arabian shores in winter. Here it is seen in its attractively speckled winter plumage; a black belly, neck and face are the features of its breeding plumage.

Above left: Another migrant wader is this Caspian Plover; photographed in spring, it wears its full breeding plumage. It occurs mostly in the north of the peninsula.

Above right: A specialized 'wader', the Collared Pratincole feeds on flying insects. Shorts legs, long pointed wings and a

deeply forked tail are all adaptations for spending the greater part of the day on the wing.

Opposite: Much sought after for food, the Tufted Guineafowl is a large gamebird found in the cultivated areas of the southern Tihama.

Opposite: The Red-billed Tropicbird inhabits all the seas surrounding Arabia, yet it is rarely seen since it lives far out to sea on the rocky islands of the Gulf and the Red Sea.

Left: Vast flocks of waders pass through Arabia in autumn on their way from as far north as Siberia to as far south as South Africa; the long journey back is made in spring. Here a large flock of Dunlin rests on a sandbank in the Gulf.

Below: A familiar seabird of the Red Sea is the White-eyed Gull though it does not appear to breed there. In winter it flies south and is replaced by other gull species migrating from the north.

Left: One of the African birds found in southwestern Arabia is the Abyssinian Roller. While at rest it displays an eccentrically long tail; in flight a vivid splash of turquoise can be seen.

Above: Some years the African Collared Dove, a breeding bird of southwestern Arabia, spreads as far north as Jeddah; its close relation, the Eurasian Collared Dove, now colonizes the northern and central parts of the country and the Gulf region. This dove is shown above.

Top right: The handsome Little Green Bee-eater is one of the more colourful and widespread of Arabian residents. It is most commonly seen on the acacia bush from which it darts to catch flying insects.

Middle right: Among the most frequently seen and colourful of the migrant species is the Yellow Wagtail. Several distinct wagtails occur in Arabia, all with different head colours. Here the black-headed is shown; other species have grey, blue, yellow or even white heads.

Bottom right: This male Rock Thrush is a member of another brightly coloured migrant family which, as its name implies, occurs most frequently in rocky areas. It can on occasion be seen sheltering in gardens, but it is usually shy and difficult to observe.

Above: Rüppell's Weaver is an Arabian bird which has African affinities. Preferring cultivated areas and acacia woodland, it constructs there its dangling nest of grasses. Here the male is seen.

Far left: The gregarious Arabian Golden Sparrow is almost entirely restricted to the Southwest, but it is usually found there in large roaming flocks.

Left: A common resident bird of the rocky areas of the central Arabian deserts is the Trumpeter Finch. It regularly congregates to drink at waterholes during the day.

Top right: The Red-breasted Wheatear is endemic to the highland Southwest of Arabia. Like most of the birds restricted to the Arabian peninsula, little is known of its habits and breeding behaviour.

Right: The seed-eating African Silverbill is a sparrow-like finch which frequents dry scrubland and gardens in the Southwest.

Far right: Some thirty-five species of warbler occur in Saudi Arabia. The identity of some can only be safely determined by examination in the hand. Here an Arabian Warbler has been trapped and ringed for study.

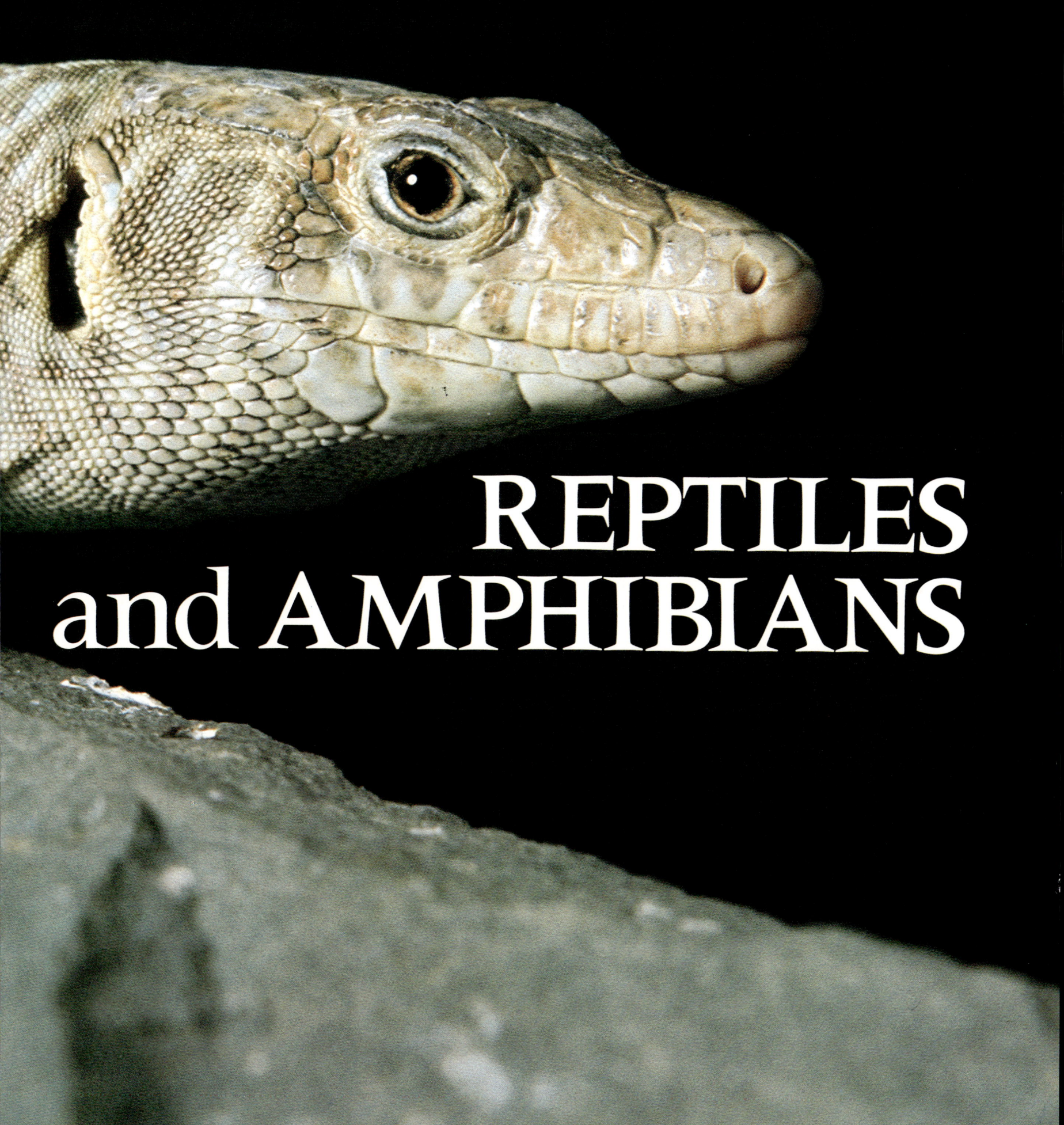

REPTILES
and AMPHIBIANS

Like many hot, dry areas, Arabia is well endowed with reptiles. About a hundred kinds of lizard and some thirty land snakes are already known to exist, and as more and more areas are investigated, new species are frequently turning up – on average one new species of lizard every year.

Reptiles are not really 'cold-blooded'; some of them may be active at temperatures well above that of the human body, even up to 45°C. They differ most obviously from the 'warm-blooded' birds and mammals in that their body heat is not manufactured internally from their food. Instead temperatures are raised by basking in the sun or by contact with objects the sun has heated. Because of this, reptiles need to eat much less than birds and mammals of equivalent sizes and, where plenty of sun is available but food sparse, as in the Arabian desert, they are often more efficient, especially as most need very little water to survive.

The most common Arabian reptiles are lizards, which can be seen in most areas, skittering through the vegetation, across dunes and along house walls. Almost all feed on smaller animals, especially insects. The various species hunt in different ways, at different times and in different places. The Grey Monitor is the largest Arabian lizard, up to 150cm (60in) long; it is forever on the hunt for food. Slender and alert with a long forked tongue that is frequently flicked out to pick up scent particles, the Monitor can sometimes be seen quartering the desert, stopping to search for smaller lizards or rodents it swallows whole.

Skinks are also active hunting lizards. The most widespread in Arabia are the Sand Skinks (two examples of which can be seen on page 50). The commonest of the three species, the Arabian Sand Skink, is superbly adapted to the problems of living on mobile desert dunes. Valvular nostrils, reduced ear openings and a countersunk jaw all help to keep sand out of the body, while running across soft dunes is made easier by fringes of pointed scales on the toes which act like snow-shoes. Other types of Arabian skink are found in moister areas, especially near the coasts. The Ocellated Skink lives secretively in leaf litter in gardens and in garbage. Several other species (one of which is illustrated on page 51) belonging to the genus *Mabuya* are more active and climb on tree trunks and rocks. An unusual characteristic of some skinks is that, unlike many other lizards, they give birth to fully-formed young instead of laying eggs.

Yet another family of active hunters are the lacertid lizards, of which three main kinds occur in Arabia. Though they lack the smooth, polished appearance of the skinks, they tend to be more elegant. Spinyfoot Lacertids, such as that illustrated on page 51, also have fringes of pointed scales on their feet which allow them to travel across soft surfaces. Desert Lacertids are found on harder ground, while the third group, Lacertas, occupy relatively moist places, and are the commonest lizard found in Europe. In Arabia they are confined to the highlands of north Oman.

Monitors, skinks and lacertids are diurnal. Most of the other Arabian lizards that share this characteristic belong to the family *Agamidae*. Among these, the dhabbs are peculiar in that they are almost entirely vegetarian when adult. Some four or five species exist in Arabia, of which the most widespread is the Small-scaled Dhabb, shown on page 53. This lives in colonies in a wide range of arid country, and swings its tail vigorously towards attackers which, since it is good to eat, are numerous and include hawks, foxes and man. Other kinds of dhabb are found around the edge of the peninsula and tend to be brightly coloured with coarser scales and often oddly shaped tails.

The other members of the agamid family are largely insect eaters. However, they do not usually hunt actively, adopting instead a 'sit and wait' strategy of perching on a vantage point and waiting for prey to pass. Seven members of the genus *Agama* are found in Arabia. Some, like the Hadramawt Agama, seen in the Southwest, are strict rock dwellers and characteristically nod their heads towards rivals. Others, such as Blanford's Agama found in the Northeast, and the Scrub Agama, prefer scrub country where they frequently sit in bushes, the males extending their strongly-coloured dewlaps.

There are also two species of Toad-headed Agamid in Arabia: the Arabian Toad-head which inhabits areas of soft sand, and the Banded Toad-head, found on gravel plains and sabkha. Both lack external ear openings, but can hear. The males frequently wag and twirl their tails, as can be seen on page 52.

By dusk all these lizards retreat into burrows and crevices and beneath stones. In warm weather these diurnal creatures are replaced by a night shift of quite different species which emerge, as twilight comes, to forage while the ground still retains the warmth of the sun. Almost all of these nocturnal lizards belong to a single family, the geckoes. The exception, Hemprich's Sand Skink (shown on page 50) which inhabits the Southwest, shares with them a vertical, slit-shaped pupil, like a cat's.

Geckoes are very varied but all have an appealing body-shape in which the head is big and rounded and the eyes large; the skin tends to be soft. Most geckoes squeak and some have distinctive calls. Like skinks and lacertids, the gecko can shed its tail if it is grasped and regrow it later. This often lets it escape from predators but it is not without cost as the tail is often a fat store and steadies the hind quarters in running.

There are over thirty species of gecko in Arabia (see pages 54-5), most of which are ground dwellers. A few are climbers and these have sophisticated adhesive pads on their toes enabling them to scale with ease even very smooth surfaces. These climbing geckoes, especially the *Hemidactylus* species, often enter houses where they sit on the walls and ceilings waiting for insects.

Though the distribution pattern of the lizards of Arabia over the peninsula is both uneven and complex, it can be roughly divided into three main regions. Desert-adapted

---

Preceding pages: A Jayakar's Lacerta looks out from its refuge in the mountains of north Oman. This lizard was first sent back to Europe in 1885 by Colonel Jayakar, a Parsee surgeon in service at the British Consulate at Muscat, and an ardent researcher after whom several Arabian animals are named.

forms usually have close relatives in the Sahara and occupy most of the peninsula; a few, like the Toad-headed Agamids have affinities with central Asian desert species. A big 'L'-shaped area, roughly from Taif to Aden and eastwards to Dhofar, contains many species with relatives in the horn of Africa, adapted to quite moist conditions. Finally, the north Oman highlands possess many lizards with relatives in the more temperate parts of Iran.

The distribution of land snakes follows the same rough pattern as that of the lizards. In most areas, however, even the dedicated observer will come across only one snake for every fifty or a hundred lizards seen. But snakes exert a fascination quite out of proportion to their abundance. Legends abound. In some areas of Arabia one species is believed to hobble camels by twining around their legs, another to launch itself like an arrow, flying 40 metres (120 feet) to strike a victim on the forehead. This last widespread legend probably refers to the Carpet Viper which may move forwards a few centimetres as it strikes.

Of the venomous land species most are vipers, which inject venom into the victim through two large hollow fangs, like hypodermic needles, at the front of the upper jaws. The fangs can be folded away. The Sand Viper, (shown on page 56), is found in a variety of dry habitats and feeds largely on lizards and rodents. It has a characteristic thick body, short tail and triangular head; some have 5mm($\frac{1}{4}$in) horn-like scales over each eye.

The two Carpet Vipers are similar to the Sand Viper. Their flank scales are diagonally arranged and toothed and produce a hissing sound when rubbed together in anger or alarm. The Common Carpet Viper is very irascible: abundant in cultivated southern steppes, it is a common cause of dangerous bites. More precipitous places are the habitat of Burton's Carpet Viper, illustrated on page 57, and, at least in northern Oman, it prefers moist wadis where it often feeds on toads. The Puff Adder, also shown on page 57, is a very stout snake and found over much of Africa and in southwestern Arabia.

Of the other deadly snakes present in the peninsula, the Arabian Cobra which can reach 200cm(80in) long, occurs in the mountains of the Southwest. Like other cobras, it spreads its neck sideways to form a hood and raises the front of its body vertically before striking. Its fangs are like those of vipers but are smaller and cannot be folded away. The extremely aggressive though seldom seen Black Cobra is confined to the Northeast. Even more secretive than this last snake is the dangerous Mole Viper, which spends most of its time beneath the soil in the moister parts.

Of the harmless or mildly poisonous land snakes most belong to the family *Colubridae*. Of these, five species of the Whip Snakes or Racers are known. Three are large, slender and highly active snakes with elegant markings: the Elegant Racer (see page 58) found in the Northwest, the Glossy-bellied Racer in the Northeast, and the Cliff Racer in the southern mountains. The latter is so named because it climbs brilliantly on rock faces.

Among other widespread colubrid snakes are the following. A common nocturnal species often found on sand dunes is the Awl-headed Snake. This has a peculiar chisel-shaped snout with which it burrows in sandy country. As can be seen on page 58, the aggressive Diadem Snake is a large creature with a bold, blotched pattern. It lives largely on rodents, whereas the Sand Snake which is about 150cm(60in) long, frequently climbs in trees and bushes where it catches birds and lizards. This snake is illustrated on page 59. The Moila Snake tends to occupy more rocky country and will distend its neck when alarmed. The mountains of the southern half of Arabia are the main habitat of the Cat Snake (illustrated on page 58). This strictly nocturnal creature feeds on birds and chameleons and other lizards. Like the previous two species, it has grooved fangs, placed at the back of the upper jaw and used to subdue small prey.

Sand Boas, small members of the same family as anacondas and pythons, are also inhabitants of Arabia. Jayakar's Sand Boa (shown on page 59) is nocturnal and typically found in loose sand; it feeds on lizards which it suffocates by coiling and tightening its body around them. A diminutive species found in cultivated areas is the Flower Pot Snake, named after one of its favourite habitats. Looking like a dry scaly worm, it is exceptional in that it is unisexual. Only females are known, and these breed parthenogenetically. The rather similar Thread Snakes resemble slender pieces of pink plastic a few centimetres long and are found in garden soil and under damp garbage.

A curious reptile related to both lizards and snakes is the Amphisbaenian, shown on page 59. Found in the Gulf region, this is a limbless creature with concentric grooves around its cylindrical body. It leads a largely subterranean existence in sandy areas, feeding on insects.

The seas that surround the peninsula are also rich in reptile life though sea snakes are found mainly in the Arabian Gulf, where fifteen species have been reported, and in the Gulf of Oman. None enter the Red Sea, but at least one kind is found along the southern coast of Arabia as far west as Aden. Related to cobras, they have similar fangs and their venom is extremely potent. Fortunately they tend to be pacific and are not very efficient in delivering their venom since, drop for drop, it is far more lethal than that of other Arabian snakes.

Marine turtles live in all the seas around Arabia and breed on its beaches and offshore islands. Both Loggerheads and Green Turtles are still relatively common though the latter is the species most usually eaten. Hawksbills, Pacific Ridleys and the massive Leathery Turtle are also inhabitants of these seas. The Hawksbill is shown on page 60.

As might be expected, Arabia has very few species associated with fresh water. A couple of terrapins, aquatic tortoise-like animals, exist and only a handful of amphibians. Of these, some of which are illustrated on pages 60 and 61, the two frogs, the four toads and a tree frog all live in the moist Southwest, although two of the toads extend eastwards to Oman and another occurs in northern Arabia, as does one of the frogs.

J.G., P.R.G. & S.I.

49

Above: Common in the moister areas of western and southwestern Arabia is this Wiegmann's Skink. Its favoured habitats include gardens and plantations.

Right: The Snake-tailed Spinyfoot Lizard is abundant in many areas of Arabia though it was only recently recognized as a distinct species. It is similar to other Spinyfoots in that it has fringes of pointed scales on its feet which aid its progress across soft sand.

Opposite top: Often seen early in the morning on soft sand dunes in many parts of Arabia is the Arabian Sand Skink. Like the monitors and dhabbs, it is esteemed as food; in the past its flesh was believed to be aphrodisiac and for this reason there was, until recently, a thriving trade in dried skinks to India.

Opposite bottom: An Arabian species found in the Southwest is Hemprich's Sand Skink. Unlike other sand skinks, but like most geckoes, it is a largely nocturnal creature, which spends the daytime in its burrow.

Top: Thomas's Dhabb is confined to southern Oman. It uses its flat, coin-like tail to block its burrow, so that pursuers are met by the hard spiny upper surface.

Above left: In the West and South, the Sinai Agama is often found in rocky areas.

The males are often coloured a brilliant bright blue.

Above right: The male Arabian Toad-head Agamid often curls and waves its tail either when it sights a lizard of the same species or when alarmed.

Opposite: The frequently seen Small-scaled Dhabb is almost entirely vegetarian when adult. It is valued as food in many areas and is consequently widely hunted.

Opposite top: Parker's Gecko is seen
mainly around the edges of the Arabian
peninsula. A climbing lizard, it often
enters houses where it feeds on insects
attracted to lights.

Opposite bottom: Carter's Semaphore
Gecko can be seen in southern Arabia.
Unlike most other geckoes, it is active by
day rather than by night and its general
behaviour is more like that of the Toad-
head Agamid.

Top right: Although it is mainly a nocturnal
hunter, the Fan-footed Gecko often basks
in the sun on rock faces by day. It has a
particularly loud call in the breeding
season.

Middle right: Slevin's Ground Gecko is one
of the commonest nocturnal lizards found
in firm sand areas where it often waves its
boldly marked tail when frightened.

Bottom right: This is one of several species
of Semaphore Gecko found in southern
Arabia. Its name comes from its habit of
flagging or waving its tail vigorously at
others of its kind.

Left: As its name suggests, the venomous and widely distributed Sand Viper is often found in dune areas. Some Sand Vipers have a long, horn-like scale over each eye.

Top: One of the most dangerous snakes in Arabia is Burton's Carpet Viper.

Its name refers to its bold body pattern, reminiscent of oriental carpet designs.

Above: The Puff Adder is found in Arabia only in the Southwest. It is not normally aggressive but it often lies on paths after dusk, and is thus liable to be stepped on when it can deliver a very serious bite.

Top: The relatively harmless slender Elegant Racer snake is confined to the Northwest of Arabia. Its bold markings include bands of black and a yellow or red stripe from between the eyes to the tail.

Above left: The Diadem Snake is a voracious rodent eater and therefore beneficial to agricultural communities. Although it is not venomous, it does bite fiercely if molested.

Above right: In the rocky areas of western and southern Arabia lives the Cat Snake, a creature that comes out at night to hunt both lizards and birds. Some Cat Snakes are marked with iridescent blue flecks.

Top right: One of Arabia's most elegant serpents is the widespread Sand Snake; its body is often striped in varying colours. It frequently climbs in bushes to catch its prey of birds and lizards.

Bottom right: Held tightly in the coils of the nocturnal Jayakar's Sand Boa, this luckless gecko is about to be suffocated and devoured.

Below: A specialized burrowing reptile, related to both lizards and snakes, is the Amphisbaenian, found around the edges of the Arabian Gulf. This 'legless lizard' habitually throws its tail forward and plants the end in the sand. It then revolves its body around this pivot, leaving a strange track in the sand.

Above: The Hawksbill Turtle is one of five kinds of marine turtle which live around the coasts of Arabia and it is the source of commercial tortoiseshell.

Left: One of only two freshwater 'turtles' to be found in Arabia, the Caspian Terrapin is common in irrigation ditches in the Northeast of the peninsula.

Opposite top: Savigny's small green Tree Frog climbs expertly in vegetation in the moist places it inhabits in western Arabia.

Opposite bottom: Often active by day, this toad (*Bufo orientalis*) lives in southwestern Arabia and northern Oman. Its weak, grunting call can frequently be heard in wet wadis.

INSECTS
and ARACHNIDS

Insects are just as varied as the rest of Arabian wildlife, because of the variety of landscape and climate. For example, there are over a hundred varieties of butterfly, innumerable spiders and scorpions – some dangerous, some extremely rare; beetles of a remarkable range of habits of life; locusts and grasshoppers.

Some species belong to Europe and temperate Asia to the north, some to tropical Africa, and some are unique to Arabia. Most of the species, including some of the truly Arabian insects, are found in and around the Asir highlands. This is because the mountains which reach a height of about 3000 metres(10,000 feet) receive higher rainfall and offer a greater range of environmental conditions.

The European and Asian ('Palaearctic') species are mostly found to the north of the peninsula: the Northwest has a Mediterranean climate. Of the creatures that flourish here many are also to be found in Europe. The beautiful Painted Lady Butterfly, illustrated on page 71, is found right across Europe as far as Britain. It is one of the regular migrants, perpetually on the move in search of good breeding conditions. By contrast, the African Lime Swallowtail, shown on page 73, is a tropical African butterfly which does not usually stray outside well-watered southwest Arabia.

Perhaps the insect most evocative of Arabia is the Desert Locust, shown on pages 62-3. It is found across that great swathe of arid territory from the western Sahara to southwest Pakistan. During recessions it occurs in the drier central part of the desert, where it survives by migrating. During plagues, however, it invades larger surrounding fertile areas, reaching East Africa, the fertile Crescent and even southern Europe. 'Invades' is the word, for until recently a plague of locusts could devastate entire regions, all green growth the victim of the locusts' appetite. Yet the locust is no different in this respect from any migratory animal – even man himself, before the coming of technology: it must migrate to where the rain has fallen and fresh vegetation is growing, or it dies.

Among insects one finds grazers, such as grasshoppers and caterpillars, decomposers which include dung beetles, carrion-flies, and wood-eating termites, and predators like the mantids, wasps, ladybirds, ground beetles and robberflies. Some insect predators do not kill or eat their prey, but lay their eggs on them. The larvae burrow into the victim and feed on the living tissues, finally killing it as they finish their own growth. These larvae are important to the farmer as they reduce the number of insect pests on his crops. Finally, there are those insects that feed on the blood of vertebrates, such as the mosquitoes and tabanid flies; these are the transmitters of disease.

Insects protect themselves against predators – both insect and vertebrate – in ingenious ways. Colour plays a decisive role, particularly among the daytime species, by providing camouflage from detection. An insect's camouflage tells about its habitat. For example, the *Sphingonotus* grasshopper, which is illustrated on page 66, could only live on stones while the *Acrida* grasshopper could only live in grass. While the *Sphingonotus'* grey-brown front wings match its

habitat, its brightly coloured hindwings signal its presence to a mate. When the insect lands on the ground, the duller front wings cover the coloured hind ones, so that a predatory lizard or bird will lose track of it.

The grasshoppers *Poekilocerus vittatus* and *arabicus*, by contrast, are so brightly coloured that they look as if they are advertising their presence (see page 67). This is just what they are doing. They feed on poisonous plants. While the grasshoppers themselves are resistant to the poison, they store it in their tissues to use against predators. The bright colouring is a warning to birds and lizards to keep off.

In Arabia many insects eat and breed throughout the year, though several species are known to hibernate during the cold season. Others are well adapted to hot and dry seasons, but are normally active during the evening or night. Some, which live under extremely arid conditions remain dormant during the hottest months of the year, a period of quiescence known as estivation. If conditions are not suitable – a continued absence of rain, for example – some butterflies will spend up to several years in the pupal stage without any perceptive signs of life; then when they finally emerge, mate and lay eggs, the next generation will reach the pupal stage in a matter of weeks to endure a further period of deep sleep perhaps lasting years.

Among those creatures that are particularly abundant in the desert because of their ability to withstand the extreme heat and drought are the arachnids. These insect predators, many of which are able to inflict uncomfortable bites and stings, include spiders, sun spiders and scorpions; some are shown on pages 74 and 75. Though they are not true insects, they are an important component of what one might call 'the insect world'.

A.S.T.

Preceding pages: The ravenous Desert Locust is a true nomad and well adapted to life in Arabia. Its swarms of millions migrate hundreds of miles in search of verdant feeding grounds where rain has recently fallen. Incredible damage is frequently done to crops by these plagues which defy control.

Opposite top: The brilliantly coloured Gold Wasp lays its eggs on other insects. When the larva hatches, it burrows its way into the host's tissues, and feeds there until the host is killed.

Opposite bottom left: The sting of the Mud-dauber Wasp is a painful one; it also has a paralyzing effect on the spiders with which it stocks its nest in order to provide its larvae with a good stock of fresh food.

Opposite bottom right: This female mosquito (*Culiseta annulata*) must have a meal of blood before its eggs can develop: here one sucks its food from a human arm. Dangerous diseases, such as malaria and yellow fever, are spread by other species of mosquito.

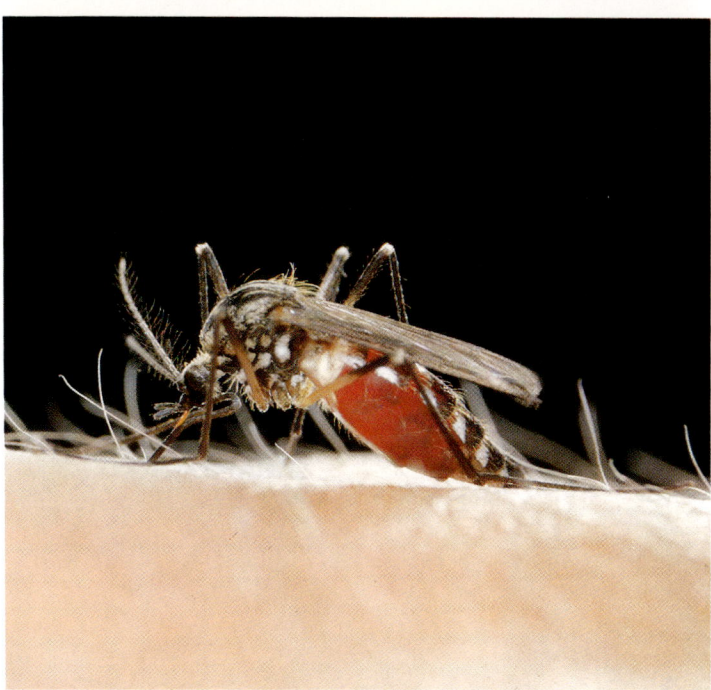

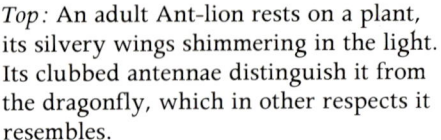

Top: An adult Ant-lion rests on a plant, its silvery wings shimmering in the light. Its clubbed antennae distinguish it from the dragonfly, which in other respects it resembles.

Above left: The dumpy sand-coloured larva of the Ant-lion employs an ingenious method of trapping food. It digs out small holes in sandy ground and waits expectantly for insects to tumble into them.

Above centre: To feed on the sap of its host plant, the Cicada pierces through its bark with specially adapted mouthparts. The

distinctive call of the male, often to be heard in hot climates throughout the world, is made by the vibration of a membrane of the abdomen.

Above right: The colouring and patchy markings of this *Sphingonotus* grasshopper enable it to be virtually invisible on stony ground. In order to match their local habitats, the different populations exhibit varying colour schemes.

Opposite top: The impressive camouflage of the *Truxalis* grasshopper enables it to avoid detection by potential predators. With its elongated shape and subtle striped

markings it merges with the grass in which it lives.

Opposite bottom left: A hunter in the long grass – a Praying Mantis stalks the desert grasslands in search of prey, using its camouflage to disguise its presence.

Opposite bottom right: Unlike many Arabian grasshoppers, *Poekilocerus vittatus*, on the left, advertises its presence with its startling yellow and black markings, a reminder to predators of its poisonous powers. *Poekilocerus arabicus*, an insect with similar habits, is on the right.

Above: This carabid beetle of the genus *Anthea* hunts other insects on the ground. Its graphic markings of a few large white spots on a shiny black body cause it to be commonly known as the Domino Beetle.

Left: Both fully-formed adults and nymphs of the *Dysdercus* species can be seen in this group of stainer bugs. Some of the most colourful members of the insect world, with bright red, black and white markings, they are pests of cotton and other crops.

Opposite: The shiny black Dung or Scarab Beetle lays its eggs on balls of animal dung, on which the larvae feed as they grow. Through this process they help to recycle the dung and thus maintain the fertility of the soil.

Opposite top: Although the oleander bush is poisonous to most animals, it is the favoured habitat of the Hawk Moth. The caterpillars feed on it and the subtlely marked adults take nectar from the flowers. Both male and female are shown here.

Opposite bottom: An American Bollworm (caterpillar of *Heliothis armigera*) fights off an approaching Chalcid Wasp which hopes to lay its eggs on it. Boldly striped and spotted with yellow, this caterpillar is a pest of cotton, tomatoes and other crops.

Top right: The delicate Bean-blue Butterfly is an African species found in the Southwest of Arabia. It feeds on many species of the pea family, but is especially fond of cultivated beans.

Bottom right: One of many regular migrant butterflies to Arabia is the beautiful Painted Lady. A species of Palaearctic origin, its range extends throughout Europe as far north as Britain.

Top: The boldly-marked *Junonia orithoyia here* Butterfly goes down in groups to the edge of water pools to drink. It is however, often harrassed by resident toads, which try to nip pieces out of its wings.

Above left: A local and rather uncommon inhabitant of the southwestern highlands is the *Acraea chilo* Butterfly. Its best protection against predators like birds is that it tastes nasty; it is one of five Arabian species of the poisonous *Acraeidae* family.

Above right: The elegant and shimmering Afrotropical Fig Blue Butterfly is found from the Asir to Oman, where its favoured food plant is the wild fig tree.

Opposite: A newly-emerged Lime Swallowtail Butterfly prepares for its maiden flight; its discarded chrysalis remains on a plant stem just below.

Above: This aggressive-looking creature is a Wolf Spider of the *Hycosidae* species. Unlike most spiders, it does not spin a web in which to catch its prey, but goes forth and hunts it instead.

Left: The Camel or Sun Spider is in fact nocturnal; emerging by night, it hunts insects on the ground. It is not a true spider, but a member of the arachnid family, the *Solifugidae*.

Despite their evil reputation and
ominous looking sting and pincers,
scorpions are loving parents and care
dutifully for their young. The sting of the
Arabian species can be extremely painful,
though it is not really dangerous.

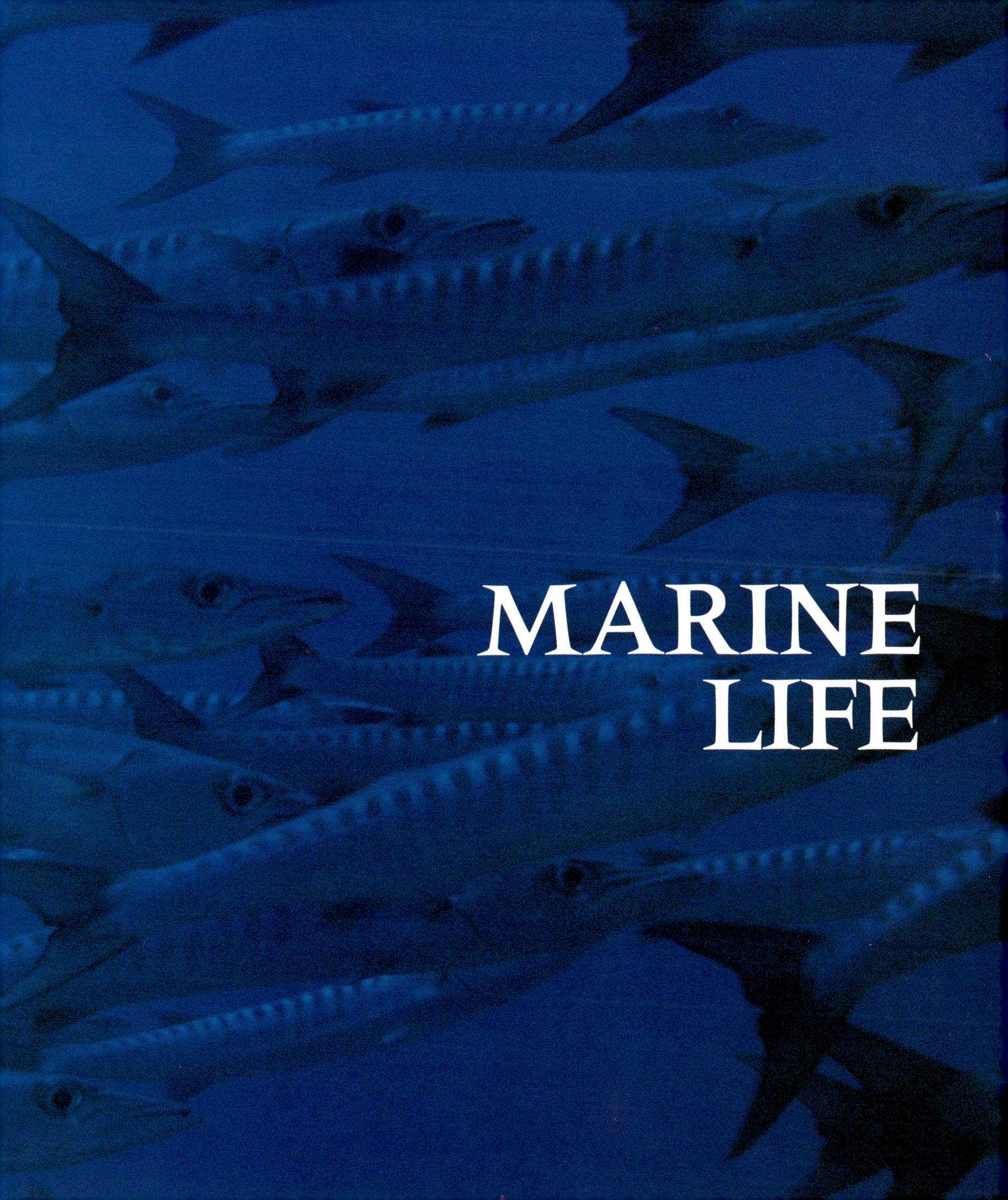

MARINE
LIFE

The reefs of the Arabian peninsula are among the most beautiful in the world. Forming a living mass of constantly changing kaleidoscopic hues, their abundant marine life is breathtaking in its amazing shapes and vibrant colours. Indeed, it is an underwater paradise so undisturbed, – except in those areas of expanding ports – that the visitor to the reefs can think of himself as a pioneer, for the areas that have been well explored are still relatively few. The most rewarding time of day to see this world is in early morning or late afternoon, when many creatures emerge from their hiding places to nibble and hunt prey.

The Red Sea is exceptionally rich in marine life being comparatively free of harmful pollutants. Although it is classified as part of the Indo-Pacific region, it is quite isolated from this larger area of water. Consequently many unique species have evolved there and it is a haven for a number of marine animals and reptiles that have elsewhere been pushed almost to the brink of extinction.

Usually separated from the shore by a shallow sandy channel, the edge of the reef flat drops away suddenly. Here on its almost sheer face are to be found many of the constantly growing coral organisms: the fire coral with its painful sting, the massive brain coral, and the delicate web-like red gorgonian coral, amongst others. The extraordinary structures of these corals provide homes and hiding-places for the varied inhabitants of the reef, which include sea urchins, anemones, starfish, slugs (such as the illustrated Nudibranch), squid, crustaceans, and molluscs as well as a wealth of tropical fish. Creating an impression of constantly changing colour and movement, they jointly make up the carefully balanced ecosystem of the reef.

Perhaps the most spectacular inhabitants of the reefs are the brilliantly coloured fish like the Butterflyfish (pages 88 and 89), Coralfish (pages 88 and 90), Clownfish (page 86), Parrotfish and Angelfish (page 87). Their often bizarre shapes are perfectly adapted to their lifestyles and habits. But their glowing colours and designs – reds, pinks, oranges, yellows, blues and greens in spotted, dotted, checked, striped or splashed patterns – constitute a complex marine riddle. Nowadays most marine zoologists believe that these lavish markings constitute, primarily, a form of sign language to indicate territorial rights and warn other individuals of the same species against entering their presence. The markings can, on the other hand, also be used to facilitate the finding of a mate in a densely populated habitat.

Many Red Sea fish are also the possessors of an extraordinary ability: that of dramatically altering their colouring. This is sometimes used to demonstrate a change of mood, as is the case with the Bigeye. Its swift transformation from bright red to pure white to mottled pink and silver is shown on page 85. More frequently this ability is used as a means of camouflage to fool predators or to remain undetected while in wait for unsuspecting prey.

Despite the many friendly and harmless fish that are to be seen, the reef's inhabitants also include some extremely dangerous creatures. Of these, some of the most poisonous make full use of their camouflaging abilities. Both the Scorpionfish (shown on page 83) and Stonefish lurk in wait for their victims, disguised as rough pieces of algae covered rock. If the Stonefish is stepped on, and one of its thirteen dorsal spines punctures a blood vessel, death can occur in an hour. The aggressive Lizardfish, which is illustrated on page 83, is capable of changing its colours in a matter of seconds to match its background.

All the marine creatures of the reef play a part in the underwater balance of life. The interdependant relationships of the different species are complex and fascinating: how does such a variety of creatures co-exist in the same environment, how is the food web structured, and how do the specialized symbiotic relationships work? One of the more extraordinary aspects of this jigsaw puzzle is the upkeep of medical health in the deep.

Several species take on the responsibility of being the 'doctors of the deep'. Two kinds of shrimp use their tiny claws to pick parasites off other creatures, while certain members of the Wrasse family will establish specific places of business where other inhabitants of the reef seek out their services (see page 90). As one fish is finished with, another moves into position, flares its fins and gills and floats motionlessly. The cleanerfish then relieves his client of both parasites and diseased tissue. Smaller cleaners will also fulfill the role of 'dentist', entering the mouths of bigger fish to 'clean' their teeth by picking out the parasites lodged there. Thus is ensured the good health of the reef.

D.M.

Preceding pages: Several types of Barracuda frequent the shallow drop-offs near the tropical reef. Fast and powerful, these gregarious predators are well equipped with a mouthful of heavy fangs which they often display as they swim.

Opposite top: The Nudibranch, a stunningly colourful marine slug, feeds on tender sponges or polyps of small anemones and corals. The polyps' stinging cells become lodged in it, providing an effective defence mechanism. Its lungs are exposed directly to the water, a unique feature that earns it its name of 'naked lung'.

Opposite bottom: One of the dangerous inhabitants of the reef is the Blue-spotted Stingray. Lying half-buried in the sand flats, it may be easily stepped on, when its sting will inject a poison that causes severe pain and inflammation.

The title of 'quick change artist' should go to the squid, for it can in an instant alter its colouring through an amazing range of kaleidoscopic hues, which, because of its bioluminescent pigment cells, actually glow.

Above: A beautiful but poisonous companion of the deep is the extraordinary Lionfish. Its spines, like hypodermic needles, inject a toxic poison.

Left: To break the hold of the Moray Eel is almost impossible. Darting out of dark holes, these powerful creatures are aggressive predators, though seldom a danger to man unless molested.

Opposite top: A master of camouflage and an aggressive predator, the Variegated Lizardfish can change its colour to match its background, in this case the mottled appearance of the table coral.

Opposite bottom: The deadly Scorpionfish, a relative of the even more poisonous Stonefish, is capable of inflicting extremely painful wounds if stepped on. Looking like a piece of weathered coral or rough rock, it will lurk in the rubble and be almost impossible to spot.

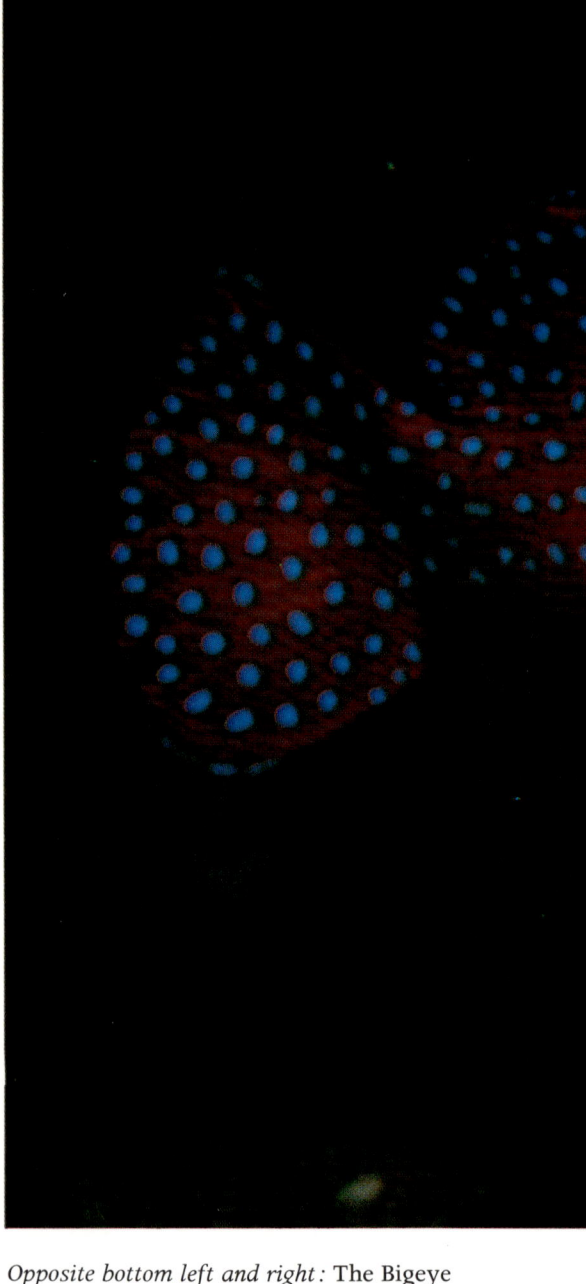

Above: One of the more plentiful and active fish of the coral reef community is the Butterfly Perch. This lovely male, seen against a background of fire coral, displays a deep-forked tail and dorsal fin.

Above right: Common throughout the Red Sea, the Coral Trout is one of the most decorative members of the grouper family. It is thought that the brilliant colours advertise the species' territory.

Opposite bottom left and right: The Bigeye must be the expert in colour change. A reflection of mood, rather than a method of camouflage, several colour alterations can take place in just a minute; in that time this individual went from bright red, to pure white and then to mottled silver and pink.

Top left: Another fish capable of startling colour transformations is the Dot-and-dash Goatfish, shown here with a cleanerfish. Skimming along the reef or sandy bottom, the Goatfish takes a taste of many morsels before deciding on his next meal.

Middle left: The graceful Batfish usually travels in schools. When young it has the same texture and colour as dead leaves and lives among sea grasses. If frightened it will sink into the vegetation.

Bottom left: The boldly marked Clownfish lives among the embracing tentacles of the large poisonous anemone, *Radianthus gelam*, to which it is immune. From this position, it will attract other small fish towards the anemone.

Opposite: The amusing Spotted Angelfish is a particularly inquisitive and friendly creature which makes it a favourite with underwater photographers and fish watchers.

Above: The attractive small Masked Butterflyfish lives in shallow coral reefs. When alarmed, it hides in crevices in the reef.

Left: A colourful and abundant member of the Butterflyfish family, the Pennant Coralfish is frequently seen exploring the reef in small groups but it seldom tolerates a close approach.

Opposite: Seen almost always in pairs or in small schools, the delightful Lemon Butterflyfish is one of the more common reef inhabitants.

Top left: The Coralfish spends most of its day nervously swimming in its well-defined territory, feeding on small organisms which it scrapes off coral and rocks.

Bottom left: A cleanerfish attends to the medical care of a grouper. Sometimes as many as a dozen large predator fish will queue up in the 'surgery', to have their parasites picked off by the 'doctor'.

Above: An extremely active fish, the multi-coloured Checkerboard Wrasse also acts as a 'doctor of the deep' and is thus a tremendous asset to the health of the reef.

Above: The Spiny Globefish is an extraordinary creature, both in appearance and in behaviour. Covered in lethal-looking prickly spines, it inflates its body to an almost spherical shape by swallowing large quantities of water when in danger.

Opposite top: The bold geometric markings in bright red and pink are the origin of the name of the Chequered Hawkfish – one of the most visually striking inhabitants of the Red Sea.

Opposite bottom: The little Blenny habitually lives in small holes in brain coral in the Red Sea, jealously guarding its home, for there is great competition for space on the reefs.

BIBLIOGRAPHY

General Publications on the Fauna of the Area
Büttiker, W & Wittmer, W (Ed), *Fauna of Saudi Arabia*, Vols I & II (1979/80, Pro Entomologia, Natural History Museum, Basle)

Buxton, P.A., *Animal Life in Deserts* (1955, Edward Arnold, London)

Cloudsley-Thompson, J.L., & Chadwick, M.J., *Life in Deserts* (1964, G. T. Foulis & Co., London)

Fozkes, H.D. (Ed), *The Natural History of Socotra and Abd el Kuzi* (1903, Liverpool Museum/Henry Young & Sons, London)

Gabriel, A.G., *Expedition to Southwest Arabia* (1954, British Museum (Natural History), London)

Hingston, R.W.G., *Nature at the Desert's Edge* (1925, H.F. & G. Witherby Ltd., London)

Philby, St. J.B., *The Empty Quarter* (1933, Constable & Co., London)

Thomas, B., *Arabia Felix* (1932, Jonathan Cape, London)

Interim reports on the results of the *Oman Flora and Fauna Surveys* (Government Advisor for the Conservation of the Environment)

The Scientific Results of the Oman Flora and Fauna Survey 1975, Vol I (1977, the Journal of Oman Studies/Ministry of Information and Culture, Sultanate of Oman; Vol II in press)

Journals of the Saudi Arabian Natural History Society (1971–80, Numbers 1–26, published from P.O. Box 5127 Jeddah, Saudi Arabia)

Mammals
Harrison, David L., *The Mammals of Arabia*, Vols I, II, III (1964/68/72) Ernest Benn, London)

Harrison, David L., *The Mammals of the Arabian Gulf* (1981, George Allen & Unwin, London)

Birds
Gallagher, M. & Woodcock, M.W., *The Birds of Oman* (1980, Quartet, London)

Jennings, M.C., *The Birds of Saudi Arabia: A check-list* (1981, Jennings Publishers, 10 Mill Lane, Whittlesford, Cambridge)

Jennings, M.C., *Birds of the Arabian Gulf* (1981, George Allen & Unwin)

Meinertzhagen, R., *Birds of Arabia* (1954, Oliver & Boyd, Edinburgh and London)

Reptiles and Amphibians
Bellairs, A. d'A., *Life of Reptiles* (1969, Weidenfeld & Nicolson, London)

Gallagher, M.D., *The Amphibians and Reptiles of Bahrain* (1971, privately published in Britain)

Goin, C.J., Goin, O.B., & Zug, G.R., *Introduction to Herpetology* (1978, Freeman, San Francisco)

Mertens, R., *World of Amphibians and Reptiles* (1960, Harrap, London)

Schmidt, K.P., & Inger, R.F., *Living Reptiles of the World* (1957, Hamish Hamilton, London)

Insects and Arachnids
Balfour-Browne, J., *Expedition to Southwest Arabia 1937–8 16. Coleoptera: Haliphdae, Dytiscidae, Byrnidae, Hydraenidae, Hydrophilidae.* (Brit. Mus. (Nat. Hist.) London 1: 179-220)

Büttiker, W. (Ed), *Fauna of Saudi Arabia* see General Publications

Larsen, T.B. & Larsen, K., *Butterflies of Oman* (1981, John Bartholomew & Sons, London for the Office of the Advisor for Conservation of the Environment of the Government of Oman)

Larsen, T.B., see General Publications: *The Scientific Results of the Oman Flora and Fauna Survey 1975* (1977)

Larsen, T.B. *see* General Publications: Büttiker, W. (Ed), *The Fauna of Saudi Arabia* (1979/80)

Pittaway, A.R., *The Butterflies and Hawkmoths of Eastern Saudi Arabia* (1979, Proceedings and Transactions of the British Entomological and Natural History Society)

Shalaby, F., *A Preliminary survey of the Insect Fauna of Saudi Arabia* (1961, Bull. Soc. Ent. Egypt. 45:222)

Marine Life
Basson, P.W., Burchard, J.E., Hardy, J.T., & Price, A.R.G., *Biotypes of the Western Arabian Gulf: Marine Life and Environments of Saudi Arabia.* (1977, Dhahran)

Botros, G.A., *Fishes of the Red Sea* (checklist; 1971, Oceanography and Marine Biology, London)

Klunzinger, C.B., *Synopsis der Fische des Rothen Meeres* (1870–1/1964, Historia Naturalis Classica/J. Cramer, Weinheim)

Neve, P., *Dangerous Sea Fishes* (1972, Journal of the Saudi Arabian Natural History Society, No. 3.)

Neve, P. & Al-Aiidi, H., *Red Sea Fish* (checklists I and II; 1972, Journal of the Saudi Arabian Natural History Society, No. 5)

Ormond, R. & Bemert, G., *Red Sea Coral Reef* (1981, London)

Relyea, K., *Inshore Fishes of the Arabian Gulf* (1981, George Allen & Unwin, London)

White, A.W. & Barwani, M.A., *Common Sea Fishes of the Arabian Gulf of Oman*, Vol I (1971, Trucial States Council)

INDEX

Illustration references are given in bold type